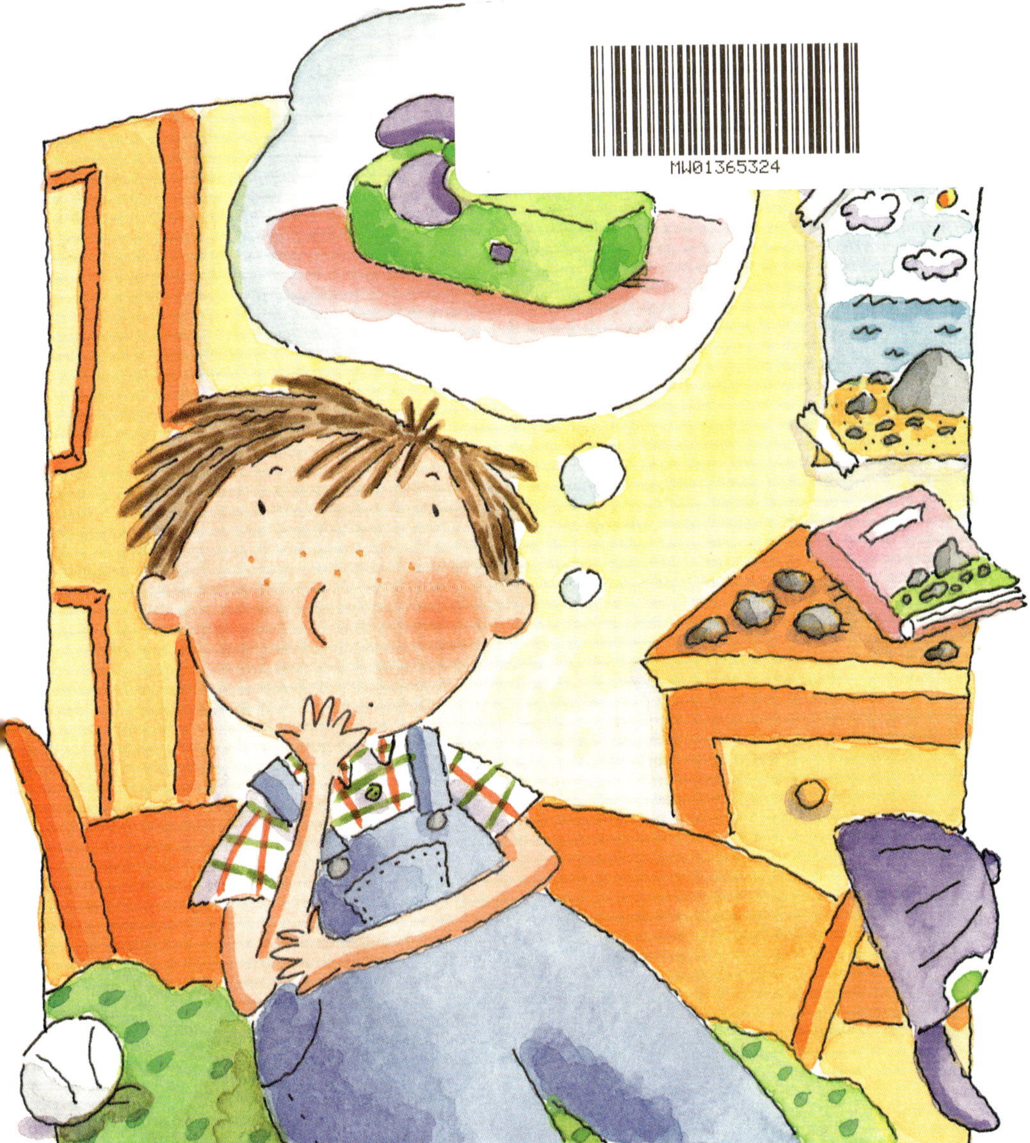

Tim's fan is lost.

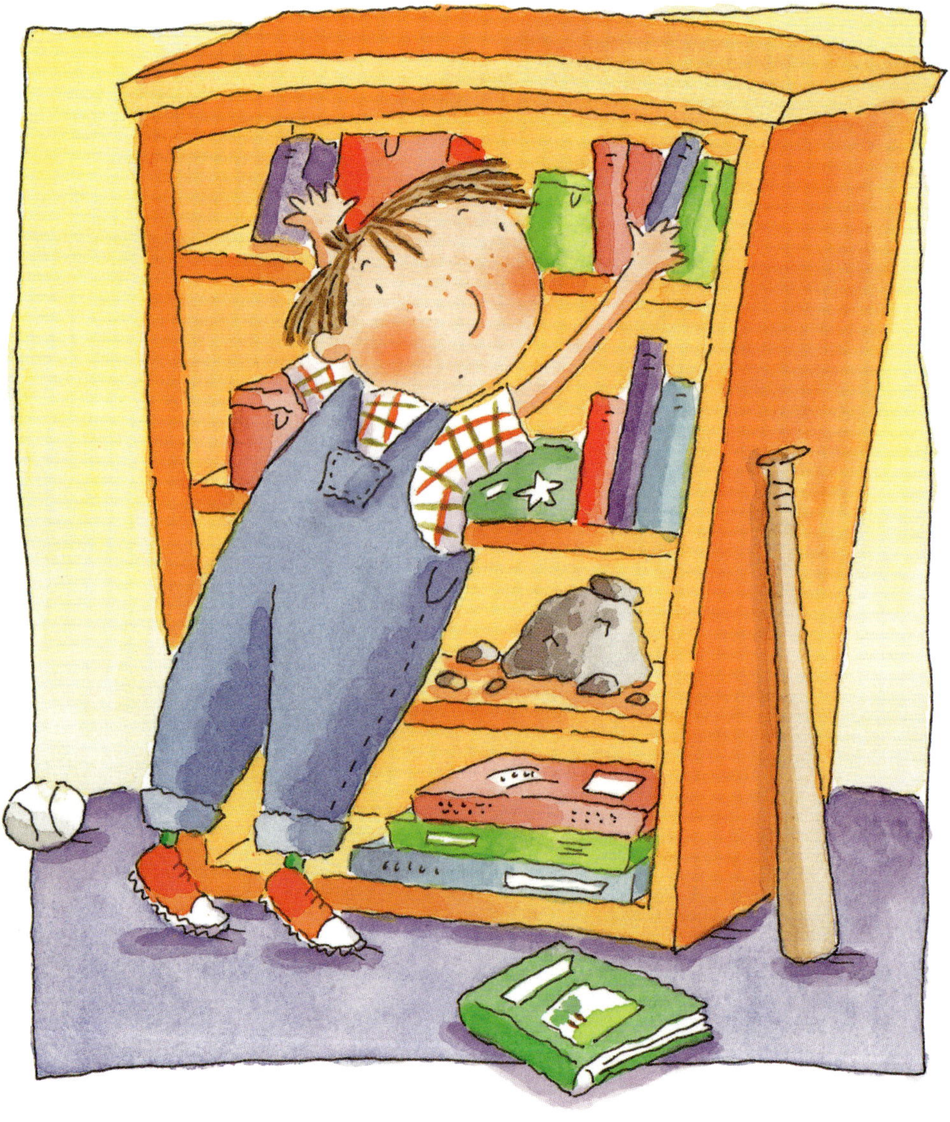

Is his fan at the top?

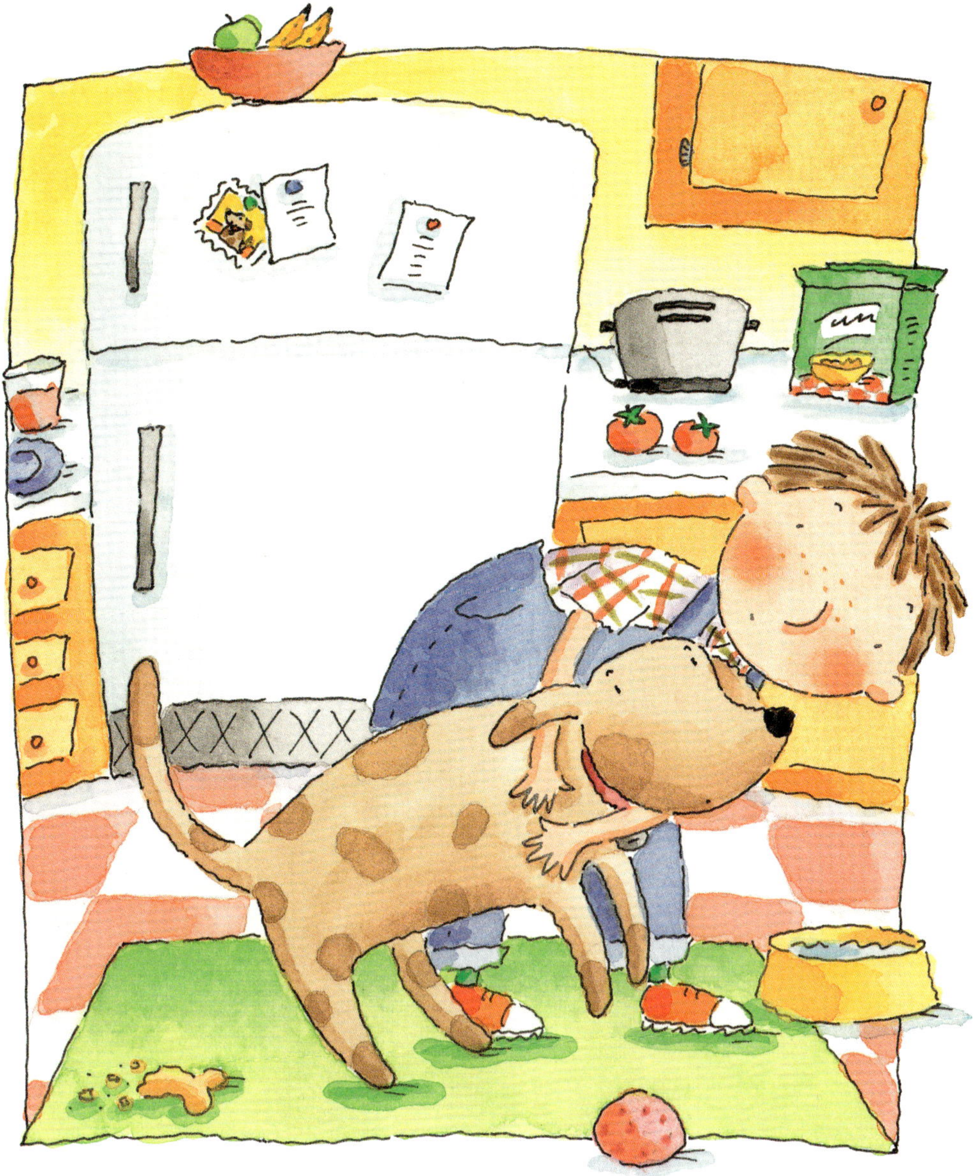

Is Tim's fan on the mat?

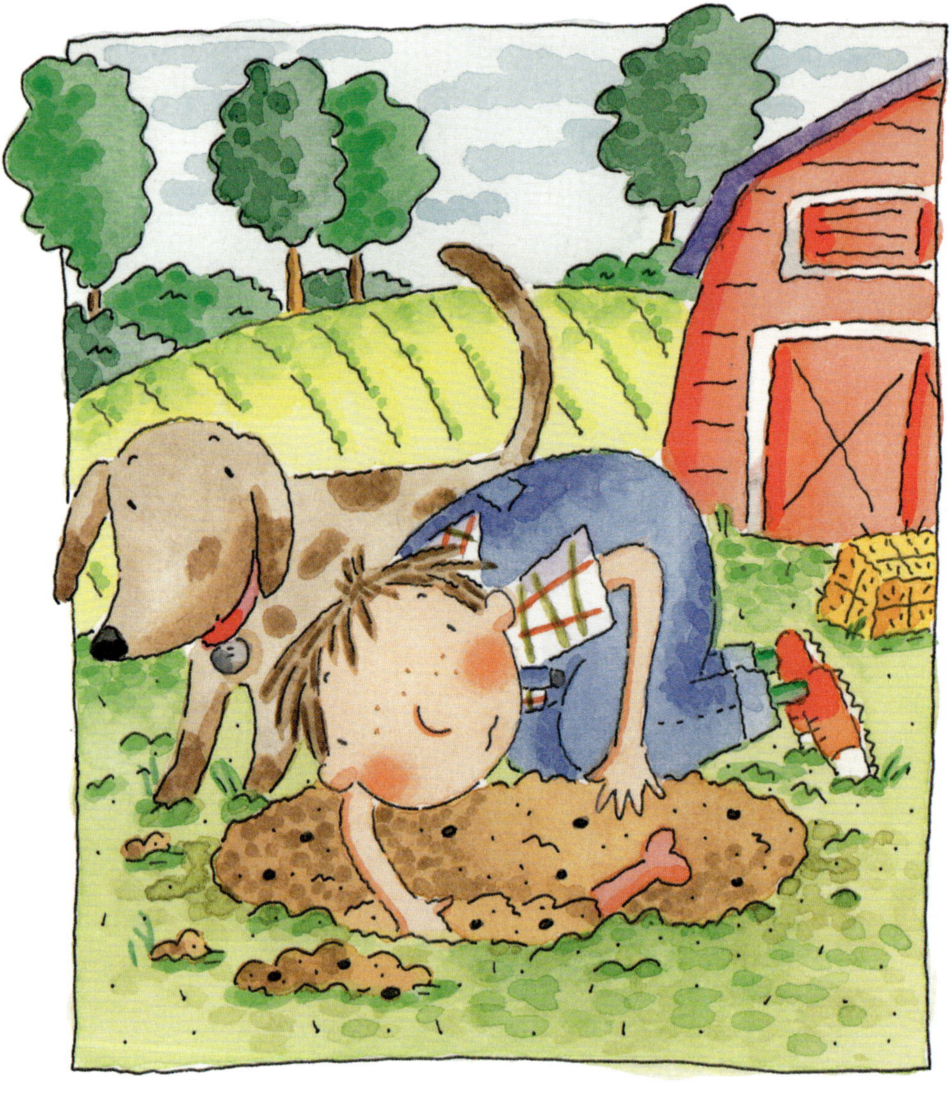

Is his fan in the pit?

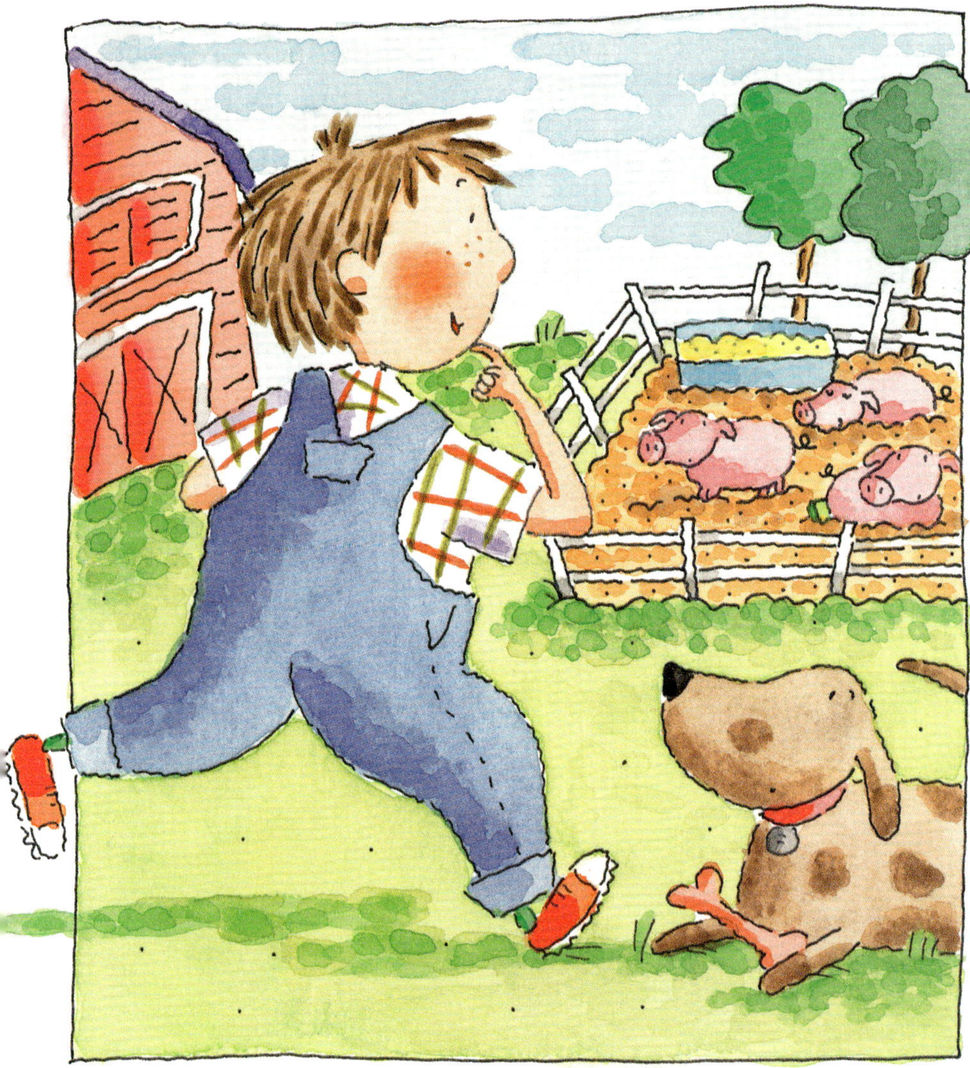

Tim's fan is **not** lost.

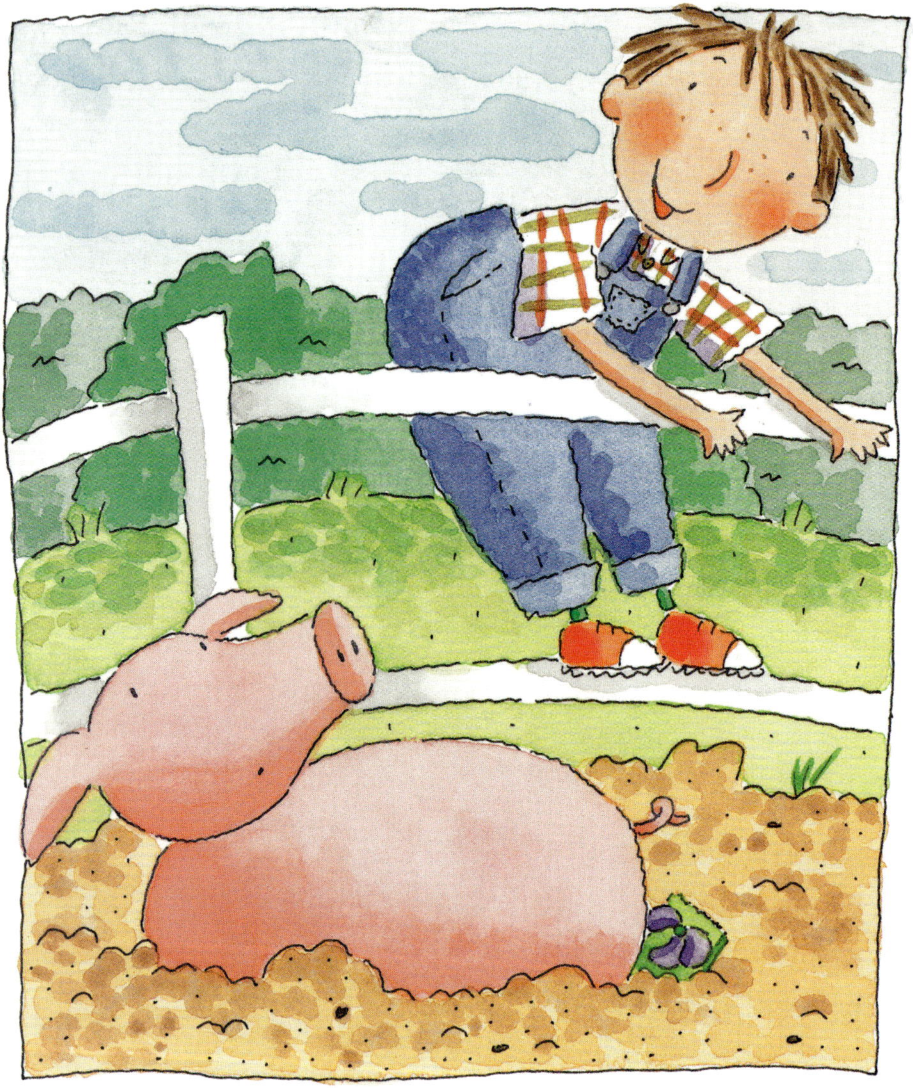

Pig sat on Tim's fan!

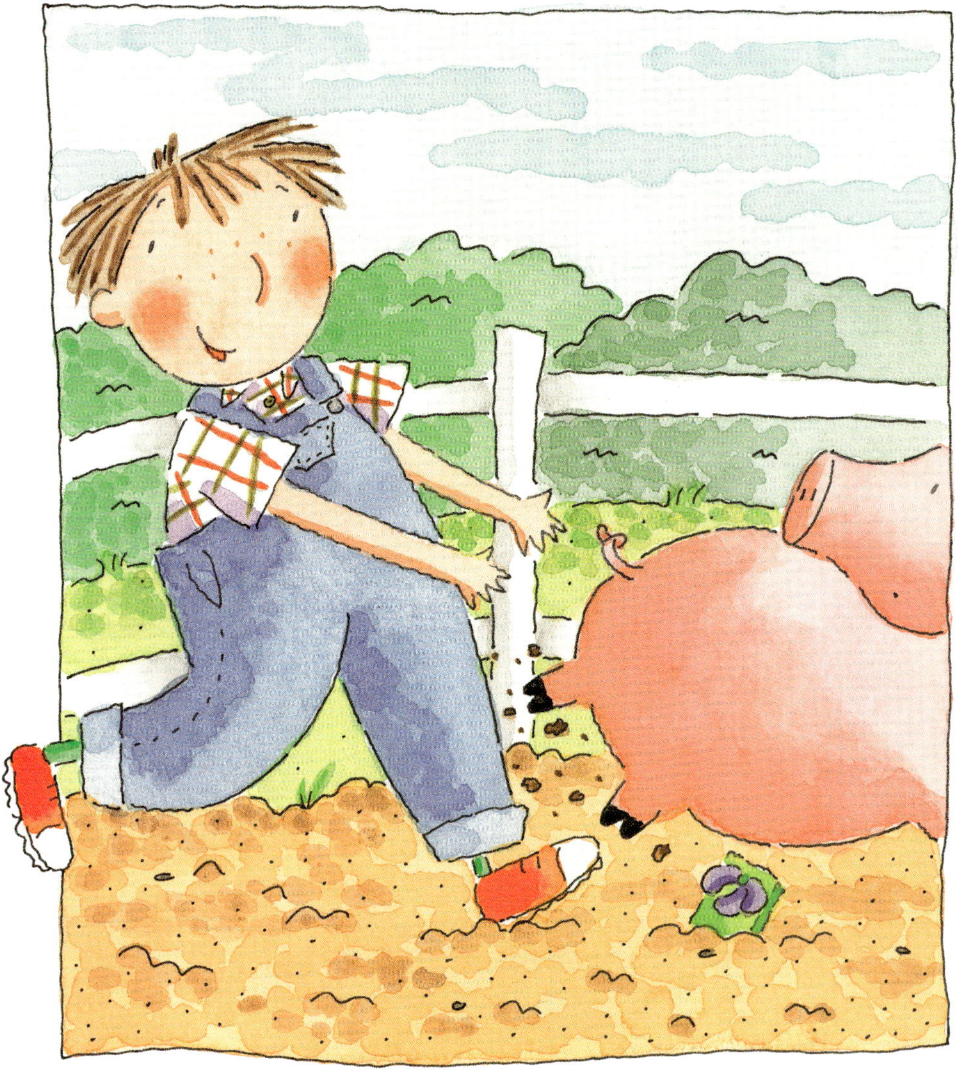

Pig got off fast.

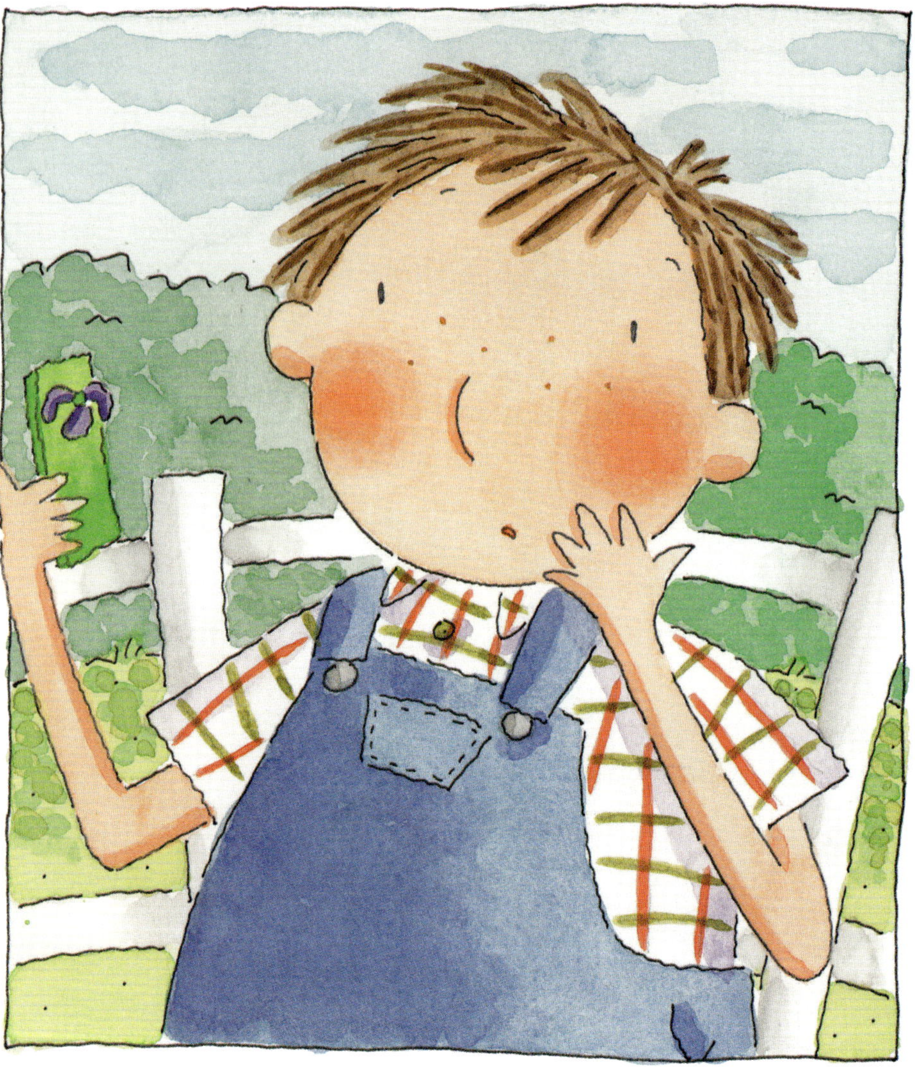

Tim's fan is flat!